Born in the north of England just before World War II. John Cowley attended grammar school from 1949 to 1957, followed by a university scholarship and teacher training at the City of Leeds Teacher Training College. He served as a commissioned officer in the Royal Navy from 1963 to 1968. John emigrated to Australia in 1968.

He has worked as a science teacher and headmaster from 1968 until retirement in 2000. Now retired, pursuing interests as a science writer, painter, and philosopher.

John holds a Master of Education (MEd) from Queensland University of Technology, awarded in 1997. He completed a Doctor of Philosophy (PhD), initially registered at Ballarat University in Victoria and completed at Calamus International University in 2002.

Dedicated to my wife, Muriel—my greatest encourager and the love of my life for over 64 wonderful years.

John Cowley

ENERGY AND EVERYTHING

AUSTIN MACAULEY PUBLISHERS
LONDON • CAMBRIDGE • NEW YORK • SHARJAH

Copyright © John Cowley 2025

The right of John Cowley to be identified as author of this work has been asserted by the author in accordance with sections 77 and 78 of the Copyright, Designs and Patents Act 1988.

All rights reserved. No part of this publication may be reproduced, stored in a retrieval system, or transmitted in any form or by any means, electronic, mechanical, photocopying, recording, or otherwise, without the prior permission of the publishers.

Any person who commits any unauthorised act in relation to this publication may be liable to criminal prosecution and civil claims for damages.

The story, experiences, and words are the author's alone.

A CIP catalogue record for this title is available from the British Library.

ISBN 9781037106200 (Paperback)
ISBN 9781037106217 (ePub e-book)

www.austinmacauley.com

First Published 2025
Austin Macauley Publishers Ltd®
1 Canada Square
Canary Wharf
London
E14 5AA

Thank you to Austin Macauley for having the faith in me to publish my book. Gratitude also goes to all the scientists and philosophers who have pioneered research into the structure of matter—without their work, this world would remain an even greater mystery. A heartfelt thank you to all the dedicated teachers: you are not only the source of knowledge but also the inspiration for future generations in the pursuit of wisdom.

Table of Contents

Earth Water Air and Fire	11
Different Forms of Energy	20
Being and Doing	24
Energy and Cosmology	26
Gravity and Time	38
In Review Time and Eternity	39
A Question of Time	43
Cosmology	45
Life and the Cosmos	46
A Few References	52
What are Stories?	54

Earth Water Air and Fire

Around 450BC, Empedocles, a Greek philosopher born in Sicily, was reputedly the first person to come up with the idea that everything we call matter consists of four basic elements earth, water, air and fire. He demonstrated that air was something rather than nothing by inverting a bowl in water and observing that the air held the water back from the bowl. Air was determined to be a separate 'element'. A hundred year later Aristotle, another famous Greek philosopher held to the same four basic elements and introduced the notion of dichotomy for example water and fire are fundamentally contrasted as fire rises and water 'falls', fire is hot, and dry, water is by nature cold and wet. Aristotle didn't attempt to explain the different states or forms of water as gas (steam) or solid (ice).

Generations of Eastern and Western philosophers from the earliest recordings of history came up with very similar ideas about the structure of matter. In a similar vein to the Greeks, yoga maintains that there are five elements of nature: earth, water, fire, air and space. Knowledge of these five elements is believed to allow the yogi to understand the laws of nature and to use yoga to obtain wisdom and happiness. Yet again the Chinese (Taoism) hold there are five elements:

wood, fire, earth, metal and water, each of them associated with certain body organs, a colour, a taste, an emotion and a season of the year. Godai (Japanese philosophy) refers to the five elements earth, water, fire, wind and void (aether). In Hinduism the five elements are fire, air, water, earth and sky.

It isn't really surprising that these early philosophers and scientists came up with very similar ideas after all they were observing and examining the same earth, water, air and fire. Empedocles was conducting experiments and then demonstrating the differences between air and water this was and still is science. There would have been some communication between East and West particularly as the various empires spread across the globe. We know from the exploits of Marco Polo and others that there was a fruitful exchange of ideas between widely separated philosophers and scientists. All of these early philosophers and scientists were restricted to what they could observe with the naked eye, and they could only hypothesize about the nature of the very small things beyond the limits of their vision. Similarly, they could only imagine what was way up in the sky above the surface of the earth. From such imaginings developed the myths and legends that still fascinate and entertain us today. Poor Icarus in his enthusiasm just got a little too close to the sun, a salutary lesson perhaps for modern space travelers to avoid getting too close to a black hole.

It wasn't until the invention of the microscope, first attributed to Zacharias Janssen in 1590, that science was able to examine the minutiae of all things classified as matter. When Robert Hooke and Antoine van Leeuwenhoek published their first observations some seventy years later, they initiated an explosion of scientific knowledge.

Observation was no longer limited to objects and intricate structures that could be seen with the naked eye. Microscopes using lenses to magnify and focus light became more and more powerful until the frequency of light waves limited their ability to resolve images. Light waves seemed to 'skip' around the tiniest objects, blurring the finer structural details of materials under observation. In 1931 Ernst Ruska is credited with developing the electron microscope which uses a high-frequency beam of electrons in place of light waves. In place of glass or plastic lenses, a series of electromagnets are used to focus the beam of electrons.

Modern science uses different terms for matter relating to its particular state solid, liquid or gas, instead of earth, water and air. Fire, rather than being an element in its own right, is seen as one example of a sudden release of energy resulting in a conflagration. Energy is also involved in the change of state of matter from solid to liquid to gas, in either direction. The energy (heat) required to change from one state into another, without any change in temperature, is referred to as latent heat. The latent heat required to change water into steam without increasing the temperature is called the latent heat of vaporisation. The latent heat of fusion involves changes between solid and liquid states of matter.

In the solid-state atoms and molecules are in relatively fixed positions and don't move from one place to another. In the liquid state atoms and molecules are free to move from place to place within a set volume which varies slightly depending on temperature. In the gaseous state atoms and molecules are free to move into the available space surrounding them depending on the energy available in the form of heat. The 'adiabatic' expansion of a gas produced by pumping it through a narrow opening into a vacuum chamber causes it to expand, using up the heat energy contained within itself and thereby reducing its temperature. Adiabatic expansion of gases, like Freon is commonly used in the process of refrigeration.

'Elements' in the Greek definition are more closely defined in modern science by their composition. Elements, like hydrogen (H) consist entirely of like units called atoms, whereas compounds like water (H_2O) contain a combination of different atoms, hydrogen atoms (H) and oxygen atoms (O) from two different elements. Atoms are the smallest units of any element whereas molecules are the naturally occurring combinations of atoms found in matter and may consist of atoms from the same element (H_2) or a combination of atoms from different elements (H_2O)

Atoms, although extremely small, can themselves be split into many different kinds of smaller particles. Three of the most common subatomic particles are called protons neutrons and electrons. Protons and neutrons are found in the nucleus, or central core of the atom, whilst electrons are found in shells surrounding the central nucleus. The combined total of protons and neutrons in the nucleus constitutes the vast majority of the mass of any atom and determines its mass number. Neutrons carry no electrical charge and are therefore neutral, as their name implies. Protons carry a positive charge and electrons which are almost massless in comparison to protons and neutrons carry a negative charge, opposite and equal to that found on the proton.

In most instances the number of electrons surrounding the nucleus is equal to the number of protons in the nucleus and both atoms and molecules are neutral, having an equal number of positive and negative charges. Ions are atoms and molecules that have an excess of positive or negative charges and are often formed when soluble elements or compounds are dissolved in solvents like water. Common salt, sodium chloride (NaCl) dissolved in water produces sodium ions with a positive charge, Na^+ and chloride ions with a negative charge, Cl^-.

The relative numbers of neutrons to protons within the nucleus of some elements can produce different isotopes of the same element. Different isotopes have a different mass number. However, because they have the same number of protons and hence the same number of electrons, their chemical properties are the same. Uranium U 238 is by far the most common isotope of uranium found when mining uranium deposits. U 238 is stable under normal conditions of temperature and pressure, unlike the much rarer U 235 isotope which is unstable and radioactive. The number attached to the element's name, 235 or 238 in this instance, corresponds to the combined number of protons and neutrons in the nucleus, the mass number. The atomic number of an element refers only to the number of protons in the nucleus of the atom. The atomic number of hydrogen (H) is 1, the atomic number of helium (He) is 2 and so it goes on upwards from the lighter elements to the heavier ones like lead (Pb) atomic number 82 and uranium (U) atomic number 92. Attempts to produce elements not found in nature are difficult because they tend to be highly unstable, radioactive and break down to form more stable elements. The synthetic element oganesson (Og) has the highest atomic number 118 and the highest mass number, 294, of all known elements.

The periodic table of elements is based on both the atomic number of an element and also the number of available electrons that can be 'given' or 'shared' in chemical reactions with other elements. Electrons are arranged in a maximum of seven shells with the number of electrons in each shell having an upper limit. Only two electrons are allowed in the first (1) and closest shell to the nucleus. The number of electrons allowed in each higher shell follows the general formula, $2n^2$ where 'n' is the shell number commencing with n=1 being the closest to the nucleus. It follows that the maximum number of electrons in each shell from the first shell (1) to the outermost shell (7) is as follows: 2, 8, 18, 32, 50, 72 and 98. Some shells are left incomplete before electrons are added to a higher shell; no known element has a complete complement of electrons in every shell.

Electrons can be knocked out of their outer shells during chemical reactions producing potential differences (PD) in electron concentrations between two points. A net flow of electrons along a conductor joining points of different potential produces an electric current, a source of energy that can be converted into other forms like heat and movement or can be stored in a battery. Conductors like metals and carbon have a molecular structure that is conducive to the flow of an electric current, whereas insulators mainly non-metals and plastics, resist the flow. Different conductors show differing degrees of conductivity, silver (Ag) and copper (Cu) being among the best conductors. Alloys, combinations of two or more different elements like steel which is an alloy of iron (Fe) and carbon (C) combine conductivity with other qualities like strength and a high melting point for use in carrying large electric currents over long distances as in power lines.

Some heavy elements like uranium 235 with large numbers of protons and neutrons in the nucleus are very unstable and break down naturally producing different forms of radiation energy, termed alpha beta and gamma radiation. Alpha radiation consists of 'heavy' particles containing two protons and two neutrons bound together, identical in structure to the nucleus of the element helium (He). Beta radiation consists of high-energy electrons which can cause skin damage similar to excessive exposure to the ultraviolet rays in sunlight. Gamma radiation is a very high-energy form of electromagnetic radiation similar to X-rays found in a different part of the spectrum from ultraviolet and visible light, which are also examples of electromagnetic radiation. Alpha radiation and gamma radiation are extremely hazardous to living organisms, the latter requiring thick shields of lead and concrete to safeguard against exposure.

Tritium, an isotope of hydrogen is the lightest radioactive element known. The nucleus of tritium has two neutrons and one proton and is produced naturally in the Earth's upper atmosphere when cosmic rays' (high-energy radiation from multiple sources in space) strike nitrogen molecules (N_2) in the air. Tritium is also a byproduct of nuclear reactors and nuclear weapon explosions. Over time the nucleus of the tritium atom breaks down (decays) producing beta radiation and helium (He) nuclei. The time it takes for half the number of atoms present in any given mass of tritium to decay is approximately 12 years, referred to as the 'half-life' of the element.

Some radioactive elements have half-lives measured in thousands of years.

The breakdown of unstable radioactive molecules like uranium 235 results in the production of high-energy neutrons in a process called nuclear fission. These high-energy neutrons then proceed to breakdown further uranium atoms producing even more neutrons in a chain reaction. Nuclear fission produces a great deal of heat which can, for example, be converted into steam, which can then be used to drive the turbine of an electric generator.

When the surface area of any piece of uranium-235 becomes too small to allow for the steady release of all the neutrons produced by fission, the pressure builds up resulting in a violent explosion.

For example, cutting a regular cube with a 1 cm side, a volume of 1 cm^3 and a surface area of 6 cm^2, into two equal pieces exposes two new surfaces each equal to 1 cm^2. This increases the surface area by 2 cm^2 without changing the volume. The volume remains at 1 cm^3, and the surface area increases to 8 cm^2.

The surface area to volume ratio is critical in many chemical reactions. Increasing the surface area of uranium-235 allows more neutrons to escape. Conversely, reducing the surface area to volume ratio constrains the release of neutrons and eventually leads to an atomic explosion. The amount of radioactive material needed to produce such an explosion is called the critical mass. The critical mass of enriched uranium 235 is approximately 47 kilograms. Other radioactive elements have a different critical mass for example the critical mass of enriched plutonium 239 is only 10 kilograms. Enriching elements like uranium can be accomplished by electron excitation using lasers or by using a series of gas centrifuges to concentrate the more active uranium 235 molecules.

What is written above is only intended as a very brief summary of the knowledge we have gained about matter over the last two and a half thousand years since the basic elements of earth, water, air and fire were first suggested by the ancient Greeks. Information has been limited to provide only what the writer considers most essential for an understanding of what is to follow. Hopefully, the above summary may also serve as a refresher for those readers who might have forgotten some 'long ago' school lessons.

Different Forms of Energy

People normally experience a drop in energy levels after prolonged exercise and their bodies respond by drawing down more energy from reserves stored in the liver and in muscles as glycogen. Glycogen is a compound of the elements carbon, hydrogen and oxygen $C_{24}H_{42}O_{21}$ produced from digested carbohydrates like glucose $C_6H_{12}O_6$ a simple sugar. Simple sugars (monosaccharides) can be converted into energy for muscle cells using oxygen

$C_6H_{12}O_6 + 6O_2 = 6CO_2 + 6H_2O$ water + energy

Simple sugars like glucose and fructose (fruit sugar) are the end products of the digestion of more complex carbohydrates like sucrose and starch. This type of energy is referred to as chemical energy; other examples of which would be burning fuels such as petroleum and coal, which release energy in the form of heat. Some of the energy released in the digestion of simple sugars is also used to produce heat and helps to maintain body temperature. Hence, we observe that one form of energy, chemical energy, can be converted into another form, heat energy.

Other forms of energy include solar, nuclear and electrical energy, frequently linked together because solar and nuclear energy can be converted into electricity in the search for more sustainable and less polluting sources of energy. Fossil fuels like wood, coal, peat and gas have traditionally supplied the main energy demands of industry, as well as being used domestically for heating, lighting and cooking, etc. Wood, coal peat oil and gas could be considered as fossilised solar energy since they are formed from the remains of dead plants that used solar energy in the process of photosynthesis. Photosynthesis uses solar energy to combine carbon dioxide from the Earth's atmosphere with water to produce simple carbohydrates like glucose and oxygen the reverse of the digestive process used by consumer organisms to provide them with energy.

$6CO_2 + 6H_2O + \text{solar energy} = C_6H_{12}O_6 + 6O_2$

Simple carbohydrates produced in the process of photosynthesis are converted into cellulose, another complex carbohydrate essential for the construction of cell walls and growth. When plants die, most of the cells completely decay, some may become submerged and compressed under overlying rock formations. Over thousands of years, these plant remains become fossilized and turn into coal. If this process occurs underwater in the absence of oxygen, conditions commonly found in bogs, then a similar product called peat is formed. Burning these fossil fuels releases the solar energy used in photosynthesis that produced the plant growth many thousands of years previously.

Still other forms of energy exist, one of the earliest forms forms used by humans, being mechanical energy, which involves movement in activities such as riding a bicycle and rowing a boat. Chemical energy stored in muscles is converted into mechanical energy by moving arms and legs and transferring energy into turning a wheel or pulling an oar.

Wind, a byproduct of solar energy, can be captured by sails to replace the chemical energy supplied by muscles. Tidal changes in estuaries, under the influence of gravity, can be converted into electrical energy by using the movement of the waves to move the blades of a turbine connected to a gearbox that turns a generator, producing electricity. It could well be argued that the energy derived from tidal movements produced by the gravitational attraction of the moon and to a lesser extent the sun, is the lowest cost and least polluting source of energy currently available. An 'energy mix' derived from solar, wind, and tidal movements is a promising alternative to fossil fuels and to alleviating genuine concerns about pollution and global warming.

Potential energy is the energy that could be released under changed conditions. Changes in the position of one body in relation to another: a rock balanced on the edge of a cliff, a large body of water held in a dam above a hydroelectric generator. A falling rock can be converted into mechanical energy crushing and heating the ground at the point of impact. Water flowing downhill under the force of gravity can be used to drive an electrical generator or turn a grinding stone to mill the grain. The mass and relative velocity of a moving body (energy of motion) determines its potential energy: potential to cause catastrophic damage when considering the possibility of large meteorites and asteroids coming into contact with the Earth.

It could be argued that defining energy by what it does, or by what it has the potential to do, doesn't provide a complete picture of energy per se. One of the basic laws of physics is that energy cannot be created or destroyed only converted from one form to another. An alternative definition of energy could be that energy is what makes 'being' possible and that being enables doing. All objects, including different forms of living things, owe their being to energy. Without energy there would be no being, nothing would exist and if nothing exists, there would be no doing; nothing would or could happen.

Being and Doing

From a human, anthropomorphic viewpoint, 'being' means being alive and belonging to one of the biological classifications of living things found in textbooks: Primates, Flowering Plants, Fungi, etc. Living things, in contrast to non-living things, can either manufacture their own food (producers), like green plants, or they are capable of stealing from or sharing food with producers. Living things that depend on producers are generally classed as either Herbivores, Carnivores, Omnivores, Parasites, or Symbionts. They demonstrate recognisable behaviour patterns which most commonly involve generating energy by consuming food, disposing of waste, growth and a variety of reproductive processes.

Interestingly, a group of organisms classified as Viruses seems to exist in a transient state, on the border between living and nonliving. The ability of some viruses to change into a crystalline form, plus their general 'piratical' reproductive methods, sets them apart from other living things. All viruses are parasitic and reproduce by usurping the normal genetic control over cell division in the body of their host. They achieve this by instructing the host's cells to reproduce viral particles instead of new host cells, replacing the host's messenger RNA (ribonucleic acid) with viral messenger RNA (mRNA). The viral mRNA changes the instructions in the process of cell reproduction in the host. Messenger RNA (mRNA) contains detailed instructions, chains of codons (units of genetic code), for the production of new cells.

Artificial intelligence combined with robotics is destined to provide some challenges for the future, especially when it comes to determining what is living rather than non-living. Humans already demonstrate a tendency to talk to communication devices like Alexa and Siri (voice controlled electronic devices) as though they were talking to real people. Sometimes it is difficult to omit the traditional please or thank you in requests and responses when communicating with such devices. Humanoid robots are already performing some of the activities normally associated with being alive and artificial intelligence is challenging human intelligence in several areas, typified by the development of chess-playing computers. Human memory is no match for the memory storage available in modern computers, both in terms of total capacity and in the speed of retrieval of stored information.

Returning to the question of defining energy per se rather than by what it does or can do. The next section of this essay will seek to find a more comprehensive understanding of energy and how this might impact cosmology, including the origin of the universe and its ultimate destiny.

Energy and Cosmology

These are exciting times for humanity as we reach out into the extremities of our own solar system with manned and robotic space flights and probe ever deeper into the vastness of space using new telescopes like the James Webb and its predecessor the Hubble. Our knowledge of the Universe increases at such an exponential rate that it is challenging to keep pace with everything 'out there'.

Increased knowledge brings about new ideas about how the universe came into being, its past and its future. The prevailing theory on the birth of the universe is the Big Bang theory, or one of its more recent variations. There is quite a deal of evidence to support the idea that the universe began as a single point and expanded to what we observe today and that it will continue to expand, possibly indefinitely, possibly not. Recent measurements of the background radiation persisting in outer space are consistent with those predicted by theoretical modelling of the explosive nature of the Big Bang theory. This explosive origin of the universe is considered by most scientists to have occurred approximately 13.7 billion years ago, resulting in a universe that is a combination of space and time commonly referred to as spacetime. All matter and energy measurable within our universe exists within the framework of spacetime and there is nothing in our universe that exists outside of spacetime.

The advent of nuclear fission and the dawn of the atomic age demonstrated that matter and energy are closely related. Einstein's classic equation defines this relationship, $e=mc^2$ where e represents the energy produced, measured in joules, m is the mass consumed, measured in kilograms and c^2 is the speed of light measured in meters per second multiplied by itself.

Because 'c' is so large, approximately 300,000,000 meters per second a vast amount of energy is equivalent to a very small amount of matter. For example, only a very small amount of uranium 235 is totally annihilated, changed into energy, when a critical mass leads to an atomic explosion. Although the atomic bomb dropped on Hiroshima involved approximately 64 kg of Uranium 235 the energy released in the explosion came from the annihilation of only about half a gram of matter, the approximate mass of a butterfly.

Nuclear fusion is the process by which energy is produced, not from the fission of larger atoms into smaller ones but from the fusion of smaller radioactive atoms into larger atoms. This is the fusion process by which 'main sequence' stars like the Sun produce solar energy. Hydrogen atoms in the isotopic form Tritium are combined to produce Helium atoms plus a large amount of energy. The Hydrogen Bomb, like the atomic bomb but far more powerful, is a thermonuclear, heat-producing device, that emulates the production of solar energy by the Sun.

Forcing hydrogen atoms to combine to produce helium requires extreme physical conditions of temperature and pressure, conditions which are reproduced by using nuclear fission as a trigger inside the Hydrogen Bomb. Achieving 'cold fusion' the joining of hydrogen atoms without these extreme physical conditions of temperature and pressure, is the holy grail of nuclear energy production. So far, cold fusion has only been achieved on a very small scale and is not considered commercially viable. Other large-scale fusion experiments have tended to require more energy input than they produce. The latest attempts at gaining a 'net energy gain' in fusion ignition, conducted at the US Lawrence Livermore National Laboratory in California, uses laser beams with what has been claimed to be 'fleeting success'.

Most stars similar to the Sun, referred to as 'main sequence' stars, have a lifecycle during which they first convert hydrogen to helium until all the hydrogen has been exhausted and then they move on to more complex fusion processes. Much larger stars use a different fusion process involving a carbon, nitrogen and oxygen cycle (CNO), a process that requires the initial conditions to be more extreme. During the formation of stars huge clouds of interstellar gas and debris collapse under the force of gravity and matter is compressed into swirling clouds that force particles to collide at a faster and faster rate, generating more and more heat as they do so. The greater the mass involved in gravitational collapse, the greater the amount of heat generated, and the temperature and pressure increase proportionately.

Some large stars burn very brightly for a short time others like the Sun reach a relatively steady state and burn for a long period of time – these are the main sequence stars. Some very large stars produce so much energy in a short space of time they explode violently releasing almost all their energy in a single instant – a Supernova.

A fuller account of the classification of stars and their life cycles is fascinating but is beyond the scope of this essay and possibly not relevant to the main theme.

Reversing the Process of Producing Nuclear Energy and putting Einstein's equation in reverse: $m=e/c^2$ where m is measured in kilograms, e is measured in Joules and c^2 is the speed of light squared it is theoretically possible to produce a small amount of matter from a very large amount of energy.

Reversing the results of the Hiroshima Bomb it would take the amount of energy released in this explosion to produce approximately half a gram of matter. The question is where could such huge quantities of energy be found and what kind of process might be involved? Gravity is seen to be fundamental to the formation of stars and the production of solar energy. It is the intense heat produced by gravitational collapse that lights the fiery furnace of nuclear fusion. Could gravity in the form of black holes also be involved in converting the huge amounts of energy back into matter?

According to the Standard Model of Physics there are four fundamental forces at work in the universe gravity, the strong nuclear force, the electromagnetic force and the weak nuclear force. The strong nuclear force is the force that holds protons and neutrons together in the nucleus of the atom. The weak nuclear force relates to the forces involved in the radioactive decay of certain nuclei and the electromagnetic force results from the interaction of charged particles. Particles with the same charge repel each other, whilst particles with opposite charges attract one another.

The strong force is carried by the gluon, a type of particle belonging to a group of five force-carrying particles referred to as 'bosons'. The electromagnetic force is carried by the photon and the weak force is carried by the 'W and Z' bosons. Theoretically, there should be a corresponding force-carrying particle of gravity, the graviton but such a particle has yet to be found. There are six different types of particles called 'leptons' that do not take part in the strong interaction and only interact via the electromagnetic and weak forces.

The total number of fundamental particles in the Standard Model is 17 but only 5 of these are force-carrying particles, or bosons: the gluon, the photon, W and Z, plus the elusive graviton. There are six different quarks that make up protons and neutrons. There are three quarks in every proton and a different combination of three quarks in every neutron. Quarks are named up quarks, down quarks, top quarks, bottom quarks, strange quarks and charm quarks. A proton consists of two up quarks and one down quark plus plus three gluons to bind them together. A neutron has two down quarks and one up quark. The remaining six fundamental particles are called leptons, they are electrons, electron neutrinos, muons, muon neutrinos, tau and tau neutrinos.

There are at least another 200 different particles that have either been found to exist or are theorized within the Standard Model. The so-called 'God particle' or Higgs Boson, is one that has stimulated great public interest because of its projected role in the formation of atoms and in seemingly keeping order in the universe. A detailed explanation of all these different particles and their corresponding antiparticles is beyond the scope of this essay and only marginally relevant to the main argument.

Returning to the relationship between energy and gravity requires some elementary consideration of the different types of field or areas of influence that surround each of the different forces previously mentioned. Direct observation of the force fields within the atom which surround the force-carrying particles is extremely difficult and involves some theoretical modeling that has yet to be fully tested. The Higgs field is a quantum field often compared to a Mexican Sombrero in which the universe is rolling around in the lowest bit of an energy surface on the dip in the brim of the hat. Don't be concerned if this idea stretches your imagination to its limits or beyond, you are among many friends.

In stark contrast, observation of the field surrounding an electromagnetic force is as easy as plotting the magnetic field around a bar magnet using iron filings, a common school science experiment. Observation of the gravitational field can readily be observed in action, as per Newton's mythical falling apples. A small telescope will reveal the gravitational field around the planet Jupiter, holding its observable moons in orbit around the planet. As a force, gravity is unrelenting in its effects on matter and escape from its influence is almost impossible. Using the thrust provided by rocket engines to counter the force of gravity $g = 9.8$ m/s^2 it is possible to escape the pull of Earth's gravity g only to be ensnared by the Sun's gravity. Escaping the Sun's gravity only leads to exposure to the gravitational forces within our galactic environment and so on. There are positions in space Lagrange points where two or more opposing gravitational forces balance each other out. There are five such points in a two-body system like the Earth/Moon system.

The unrelenting nature of gravity is exemplified in the 'inverse square rule' used to calculate the change in gravitational forces with changes in distance between different large bodies of matter. In simple terms the force of gravity diminishes the further you are from the source by a factor involving squaring the proportional increase, double the distance and the force is quartered, triple the distance and the force is reduced to one-ninth. It is interesting to note that all forms of radiation follow the same inverse square rule.

Applying the inverse square rule, it can be seen that the force of gravity rapidly decreases with increasing distance between bodies but never reaches absolute zero. Reversing this relationship illustrates how gravity increases rapidly as the distance reduces between bodies in space, halving the distance between bodies increases the force of gravity fourfold, halving the distance again increases the force of gravity by a factor of 16 and so on, very rapidly. This is why in getting closer and closer to a black hole, one eventually reaches a point where nothing can escape, the event horizon. At this point the force of gravity cannot be overcome by any other opposing force. However, what happens if this continues until matter is crushed entirely out of existence and one arrives at a dimensionless point, a 'singularity', is a matter of much speculation.

There is as yet no evidence of a force-carrying particle, a graviton, to complete the number of bosons predicted by the Standard Model. Additionally, there is no evidence to support the concept of antigravity and a corresponding antiparticle to the graviton. There may be another explanation for the force of gravity that does not involve gravitons. It may even eventuate that gravity cannot be considered a force in the same context as the other four bosons: electromagnetic, strong, weak and W and Z, which form part of the Standard Model. Some scientists have suggested that gravity is better understood within the context of quantum mechanics.

Quantum mechanics had quite a controversial beginning, pitting the conflicting ideas of two great scientists against each other. In 1927, Einstein made the famous statement that he did not believe "God played chance with the universe". Chance being inherent in some of the ideas being put forward by Neils Bohr and Max Planck: ideas (quantum theory) which later developed into the science of quantum mechanics. A hundred years later it is generally accepted that quantum theory is one of the most successful theories in the history of science. This is mainly because the theory has been highly successful in predicting the outcomes of several recent experiments conducted in particle physics. Many of these experiments have involved high-energy collisions between atomic particles in a large cyclotron a particle accelerator like the 'Large Hadron Collider' located at CERN on the Swiss border with France. Using superconducting magnets along a 27-kilometer circular track, groups of protons or alpha particles are accelerated to speeds close to the speed of light and then they are used to collide with other particles on a screen. The results scattering of particles resulting from such collisions, have led to the discovery of over 200 different subatomic-particles and there may well be more to follow.

Max Planck is considered to be the founder of quantum physics in so much as he was the first to use the term 'quanta' in relation to energy. He developed the idea that energy comes in discrete parcels in bits rather than as a continuous stream (analogue) Human experience leads one to think of energy in terms of waves, feeling the waves of heat coming from the Sun and warming the Earth. Surfers riding waves on the surface of the sea transfer the energy of the waves into the movement of surfboard and surfer. The waves of radiation from a star can also behave like streams of particles – photons that can be bent by the gravitational attraction of other large bodies in space, producing the gravitational lensing effect observed in astronomy. Gravitational lensing allows astronomers to see very faint objects that are normally too dim even for the best telescopes, by bending and focusing the light emanating from them into a narrower and brighter beam.

The dual nature of light, the ability to behave as waves and also as particles, has been a subject of scientific research for over two hundred years. A light wave typically has a peak, a high point and a trough, a low point. The difference between the peak and the trough is the amplitude of the wave and corresponds to the brightness of a light source. When peaks from different light sources combine, they increase the brightness; when a peak and a trough coincide, they reduce the brightness by cancelling each other. The frequency of light waves, the number of waves per unit of time, relates to a classification of light waves into seven categories, from lower to higher frequency: radio waves, microwaves, infrared, visible light, ultraviolet rays, X-rays, and gamma rays. The visible spectrum is further divided into seven colours: red, orange, yellow, green, blue, indigo, and violet.

Photons only behave like particles when they are moving, and they have zero rest mass. The classic physics 'Double Slit' experiment was first devised by Thomas Young in the early 19th century. In this experiment, a coherent beam of light is passed through two narrow, parallel slits in a screen and projected onto a second screen placed behind the screen with the slits. Light waves are bent by diffraction at the edges of each narrow-slit projecting images on the second screen consisting of a pattern of light and dark bands. Light bands occur where the waves of light passing through separate slits combine and amplify the light falling on the screen and dark bands occur where the waves of light cancel each other out and no light reaches the screen.

Coherent beams of light consist of light of a single color and frequency or laser light in which all the waves in the beam of light are synchronised so that peaks and troughs coincide exactly. Diffraction, or bending of light as it passes between narrow slits, can be observed by allowing a beam of light to stream between lightly closed fingers and observing the blurring effect at the edges of the fingers where tiny gaps allow the beam of light to pass through.

The 'double slit' experiment has been repeated many thousands of times with variations in slit sizes, distances between slits and a variety of light sources. It is frequently quoted as proof of the duality of light by demonstrating that light can be seen to be behaving as waves. However, the experiment also produces some seemingly strange results when the beam of light passing through the slits is reduced from a stream of multiple photons to single photons, one at a time. Single photons appear to be able to cancel each other out, even when a second photon arrives later in time than the first photon, a retroactive phenomenon contradicting a long-held principle that 'time is irreversible'.

Other double slit experiments have been conducted along similar lines using electrons in place of photons, with the same results, suggesting that all particles are capable of behaving as waves. It is highly possible that this behaviour is not restricted to subatomic particles and even larger compound objects may have a corresponding 'wave function'. Louis de Broglie, a French theoretical physicist, is credited with first suggesting in 1924 that all matter has an associated matter wave. In 1933, Irwin Schrödinger, an Austrian physicist, won the Nobel Prize for Physics building on de Broglie's 'matter waves' and developing the idea of quantum wave superposition and collapse, the so-called 'observer effect'.

The Heisenberg Uncertainty Principle, proposed by the German physicist Werner Karl Heisenberg in 1927, concludes that the very act of observing alters the position of the particle being observed and makes it impossible to accurately predict its behaviour. Consequently, it is not possible to measure both the momentum and the position of a particle simultaneously. This observer effect on the activity of particles like photons can be likened to the effect any player in a game of football has on the movement of the ball. Any player not directly in control of the ball can only guess where the ball will move next because their presence will affect the decisions of the player in control of the ball. In fact, everyone on the field can be considered to be exerting influence over the movement of the ball because they are in the 'field of play'. Conscious and subconscious decisions overlapping in the mind of the player in control of the ball, combined with the chance effects created by ground conditions and weather, make exact predictions impossible.

It may seem very strange to suggest that photons are similarly 'conscious' of the presence of the observer, or alternatively that the mind of the observer is somehow controlling the movement of the photon. One possible explanation of the 'observer effect' could involve the way the frequency of light changes as it travels from source to observer. Over very large astronomical distances, the frequency of light waves is observed to decrease with increasing distance, producing what is called the 'red shift', a shift in the visible spectrum towards the red end. If light waves can be stretched in this way, then photons probably have the same elastic properties.

This 'red shift' in the frequency of a light wave can be illustrated using a piece of elastic material with tiny black marks placed very close together at regular intervals along its length. The more the elastic stretches, the further the black marks move apart, and the black seems to fade into ever fainter shades of grey. Applying this illustration of the red shift suggests that photons can be stretched, and their actual position becomes a question of probability rather than certainty. Observation leads to a superposition of wave functions of photon and observer thus fixing the photon's position in relation to the observer, whilst rendering any measurement of its momentum impossible.

Gravity and Time

Gravity and time share something in common when it comes to the difficulty of providing a definition that doesn't rely entirely on what they 'do', rather than what they are. The best definition of time is possibly, "time is what prevents everything happening at once". Time separates events, allowing space for 'everything' to exist in its own time. Gravity, on the other hand, pulls 'everything' that exists together until, in the case of black holes and singularities, there is no time and no space. By referring to Spacetime as Time-space, we give order to the sequence of events following the 'big bang'. The first requirement is for time to create space for everything else to happen, space for being and doing. Time and gravity could be considered as opposite forces involved in the process of creating and recycling universes through a successive series of big bangs, black holes and singularities.

The universe could be compared to a 'cosmic ballet', choreographed and orchestrated by the opposing forces of time and gravity. Everything and everyone within it dance to a timely tune until the curtain falls and the stage becomes quiet, awaiting the next performance. Time continues to conduct the orchestra until gravity brings down the curtain.

In Review Time and Eternity

It could be argued that if the Universe had a beginning, the Big Bang, then according to the best definition of time, it must also have an end. William Thomson, Lord Kelvin, professor of Natural Philosophy at Glasgow University in the late 19th century, is credited as being the first to suggest that the Universe will end in heat death. According to Kelvin's definition of heat death, the Universe will continue to expand until the temperature is close to absolute zero degrees Kelvin or -273°C and there is no more heat energy left. However, if we adhere to the principle that energy cannot be created or destroyed only changed from one form to another, then the idea that the Universe will end in this way is questionable.

An alternative possibility is that the beginning of the Universe, the Big Bang, was not an isolated event but rather a recycling of a previous universe, which was itself a recycling of a previous universe and so on, ad infinitum. Each incarnation of Spacetime, however long it lasts, has a limited lifespan – one story with a beginning and an end. One story out of a book full of a countless number of different stories, each with its own beginning and end.

For this latter possibility to be remotely credible, there would need to be some form of recycling mechanism operating within each successive universe. A recycling mechanism whereby everything is reduced to nothing and from nothing emerges a new version of Spacetime. One potential candidate would be a supermassive black hole, similar to the one found at the centre of Earth's home galaxy, the Milky Way. Some black holes are large enough to use the force of gravity to reduce an area of Spacetime (everything) within its event horizon to a singularity.

Such a singularity could be the point source of a future Big Bang for some, yet-to-be-determined reason, the amount of collapsed Spacetime reaches a critical level. Recycling may occur periodically, or even synchronously, across the entire cosmos, as a number of singularities reach the same 'pregnant' stage ready to birth the next incarnation of new universes.

Some black holes are thought to have formed from the remains of supernovae, giant stars that eventually burn so brightly they can no longer contain the energy they produce and consequently explode with a massive outpouring of energy. Following this supernova, what remains of the original mass of the star, collapses under its own gravity to form a black hole. The motion of stars, and other bodies within the galaxy eventually brings some of them close enough to the newly formed black hole for gravitational attraction to cause a merging, or submerging, of matter, resulting in an even larger black hole. Eventually, the black hole will reach a point at which the gravitational force produces a singularity at its centre and Spacetime is reduced to an infinitesimally small point, a singularity.

According to Einstein's 1915 Theory of General Relativity, a singularity describes the centre of a black hole, a point of infinite density and gravity within which no object inside can ever escape, not even light. Once again, we have a definition that describes what a singularity does in terms of reducing and constraining matter to an infinitely small point. Speculating about what happens within a singularity could be considered pointless (no pun intended), particularly because, from Einstein's description, it may not even have an inside. Describing what is likely to happen to objects that enter the event horizon (point of no escape) surrounding a black hole could provide some clues about the nature of a singularity. As objects enter the event horizon, they begin to be stretched, because the force of gravity is greater on the first part of the object to enter than on the latter parts, often referred to as 'spaghettification'. As the force of gravity continues to intensify, the spaces between molecules and between atoms are reduced, increasing the density of the object.

It is probable that the next phase of gravitational contraction involves a reduction of the space between the nucleus of the atom and the shells of electrons surrounding it. Still further gravitational contraction would result in protons and neutrons being pushed closer and closer together within the nucleus, eventually leading to the formation of a singularity. The gluons that bind the protons are squeezed out of position, exposing the strong repulsive forces between these positively charged particles. Protons and neutrons eventually break down into their respective numbers of up quarks and down quarks and gluons (strong forces) are released. The relative numbers of up quarks and down quarks released from a proton (2 up, 1 down) and neutron (1 up, 2 down) tend to cancel each other out, annihilating mass (protons and neutrons together account for almost all the mass of an atom). Stretching photons reduces the frequency of the electromagnetic waves and the increasing force of gravity squeezes and reduces the amplitude. Light, in fact the whole electromagnetic spectrum, is flatlining, being squeezed out of existence.

The above suggested sequence of events doesn't attempt to explain what happens to other particles trapped within the singularity. Mass may be reduced to almost zero but there is now an increasing amount of locked-up energy – potential force: like a coiled spring, waiting to be released. What possibly could trigger this release involves further speculation. It may be something as straightforward as reaching a limit, a point when the force of gravity can no longer withstand the opposing forces building up within the singularity, resulting in the singularity exploding in another 'big bang'. Possibly, the loss of mass towards the the centre of the black hole, during the breakdown of protons and neutrons into quarks, produces a corresponding sudden drop in the gravitational force.

A Question of Time

As previously suggested, the best definition of time is "time is what prevents everything from happening at once". Time is measured as an interval between the beginning and the end of any defined event, such as the rising and setting of the sun. A standardised time scale is used for comparing the lengths of different events. An atomic clock, which measures the interval between consecutive pulses from a radioactive source, is used where the intervals of time are very short. One second is the time interval between cycles of radiation produced by the transition between two levels of the element cesium 137. Seconds are scaled up to provide larger units; minutes, hours, years etc, and scaled down to tiny fractions of a second. A yactosecond is theoretically the smallest unit of time, equivalent to one septillionth of one second. The shortest span of time ever recorded in science is 247 zeptoseconds, the time it takes for a single photon to pass through a molecule of hydrogen. A zeptosecond is one quadrillionth of a second.

Measuring time intervals on a larger scale, as in the time it takes light to reach the Earth from the nearest star, Proxima Centauri, requires the use of a much larger unit the 'lightyear' which is the distance travelled by light in one Earth year. On this scale, the nearest star, Proxima Centauri, is approximately 4.25 light years distant from the Earth. Using a 'light year' as a unit of distance gives astronomers a more practical unit to measure the vast distances involved in examining the structure of our universe.

Defining time as an interval between events raises questions about what is meant by the word, eternal. A common dictionary definition of eternal is, "time without a beginning and without an end". If there is no beginning, then nothing would ever come into being and consequently there is nothing to come to an end. Possibly a better definition of eternal is one that completely excludes time and so it would follow without space, outside of Spacetime as we know it and hence outside of this observable universe.

If this current universe is a recycled version of a previous universe, in a perpetual chain of events which has no beginning and no end, then eternity can be likened to Shakespeare's empty stage, a blank Turner canvas, or the pages on which a Dickensian novel is about to be written. This proverbial 'pregnant pause' is awaiting inspiration, waiting on a beginning, waiting on time.

Cosmology

One of the most interesting things about the study of cosmology, from the writer's perspective, is that it brings together ideas from several different disciplines. Perhaps the most obvious one is science, and the various laws and precepts involved in the study of physics, chemistry and biology etc. However, the study of cosmology also involves philosophy, the study of the fundamental nature of knowledge, reality and existence. Cosmology also involves religion, the way people deal with ultimate concerns about their lives and their fate after death. Rather than being an exclusive area of study only for academics, cosmology is an area of human interest that invites comment and speculation from a wide range of perspectives. Some of the most exciting ideas and possibilities about the cosmos have been generated by a genre of books classified as science fiction and films about space exploration and adventure. Authors like Arthur C. Clarke, Isaac Asimov and Steven Spielberg, amongst many others, have generated a widespread interest in cosmology and the many questions it raises about human existence and the future.

Life and the Cosmos

Around 1600, Zacharias Janssen, a Dutch spectacle maker, is credited with making the first compound microscope using a combination of two lenses to examine objects too small to resolve with the naked eye. This opened a whole new world of living organisms and provided much greater detail about the structure of matter in general. Since this time, scientists have immersed themselves in the search for fundamental truths about the universe by breaking down the complex to find the basic building blocks of matter and nature. In many instances, this search was motivated by a desire to find answers to philosophical questions about the origin and meaning of life. Many aboriginal myths and legends illustrate how humans have pondered such questions for many millennia. One 'Dreamtime' story, about a volcano, 'Budj Bim', still being told to this day by Australia's First Nation People, relates to traceable geological events which occurred some 37,000 years ago.

Having previously drawn several comparisons between the unfolding of cosmic events and the turning of pages in a book, it may be worthwhile pursuing this analogy even further. Instead of atoms and molecules combining to make compounds, we could substitute letters of the alphabet combining together and forming words. Words arranged in sentences convey information about various 'beings and doings' and can then be formed into like groups or paragraphs. Chapters break up a story into parts to provide logical structure to a long and complex narrative. This analogy may be challenging enough for some readers but what follows is probably even more so.

One rather radical suggestion would be that the story is not just a convenient analogy, an aid to understanding science but that the reverse is true. Science is an aid to understanding the story. The earlier part of this story of the universe/cosmos, which is all about 'energy and everything' was already in existence before it was 'discovered' by science. Everything from the macroscopic – galaxies, stars, black holes, singularities down to molecules, atoms, protons, neutrons, electrons, quarks etc. must have existed well before scientists came along to 'discover' them. In the same vein, Columbus didn't discover America; it was already in existence and had been for Millenia.

The story about 'energy and everything' is essentially a story about life, a story that science has sought to unravel in part through the examination and analysis of the minutiae of everything. Stories can be written in many different ways using either the language of science, philosophy, religion or any combination the author chooses. Stories can be augmented by illustrations, told entirely in pictures, or expressed in various art forms.

Long stories can be developed into a book, or a series of books, depending on the author's intention in terms of subject, scope and intended recipients. The book, or books, could be made into plays or films with living players and become so realistic that it would be difficult to distinguish between the story and what we think of as 'real life'. I believe William Shakespeare first suggested this idea in his play, *As You Like It*:

"All the world's a stage and all the men and women merely players. They have their exits and their entrances; and one man in his time plays many parts".

The Greek mathematician and philosopher Pythagoras similarly suggested that this world was like a stage/Whereon many play their part, the lookers-on, the sage.

On the 'larger stage' the author is neither Shakespeare, nor is it Pythagoras, although both had a part to play. Some would believe that there never was an author as such, and it was/is all down to chance. Whatever 'hand' wrote such a story as this, it began a long time ago, billions of years before humanity appeared on the scene.

To set such a stage and cast all the players thereon appears beyond human genius and yet the players are no mere puppets but contributors to the story, each in their own unique way. Every bit part, every player, is essential to the bigger picture, the never-ending stories within stories, written over and over, again and again, with each successive incarnation of the universe. Not a single 'Groundhog Day' but every story unique and yet, somehow, with a familiar thread linking them together into an even bigger story, an even bigger picture.

'In the beginning was the Word and...'

Philosophers and theologians have pondered at great length over the purpose behind these stories and many have even considered the possibility that there was/is no purpose at all. However, even the most ardent atheists, like Richard Dawkins, find it difficult to silence the voices of those who believe in an author/creator. Like Charles Darwin 200 years previously, Richard Dawkins in his book, *The Selfish Gene*, presents a powerful argument for the mechanism of evolution, the driving force behind life and the rise and fall of different species. Despite this, many questions remain, such as why genes, where did they originate, how did the helical structure of DNA develop and above all else, for what ultimate purpose. And so, the questions go on and on.

It is truly exciting to be a player, however small one's part, however brief one's time on stage may be; being in the story now is what really matters. The future of everything may, or may not, be predetermined. The players may themselves be contributing to the future, or multiple possible futures. Applying Heisenberg's uncertainty principle to the actions of all the players on the stage of life illustrates how unpredictable the future can be. Students of biology are well aware of the 'web of life', wherein everything is interconnected in some way; consumers rely on producers and they both rely on decomposers to regenerate the essential elements of life for use by successive generations. Applying this same principle to a 'solar web of life' around our sun, or wider still to a 'galactic web' around the 'Milky Way', or ultimately a 'cosmic web' around our universe, may not be so very far from reality. The choices each individual, each society makes may have tenuous links to the wider world, but they are nevertheless real, as history has so frequently taught us. Perhaps it is no accident that the distances in space are so vast between planets, between stars and galaxies when comparing them to geographical spheres of human influence on planet Earth.

Humanity is already facing many challenges, non more pressing it would seem than the potential for self-annihilation through unresolved global conflicts. As the dominant species on the planet Earth, humans tend to prioritise those things which promote their own selfish interests above all other considerations – the selfish gene? This is despite the fact that such self-interests may be detrimental in the long term to our own survival. The human species has tended to neglect many of the other species that share planet Earth, unless of course, they provide food, or pleasure, or otherwise serve humanity in some way. Having belatedly come to realise how such selfish behaviour has resulted in the depletion of some species and the extinction of others, some limited plans are underway to preserve the diversity of life – Unfortunately when sacrifices are required to enact such plans, enthusiasm seems to wane and is frequently replaced by apathy, or is this just another example of the 'selfish gene' at work?

Humanity is on the verge of exporting all its deviant behaviour into the wider reaches of the solar system without having corrected some of its most severe manifestations. Slaughtering each other because we can't agree about who owns the land, causing famine through the overindulgence of some at the expense of others and carelessly, or deliberately, spreading disease as acts of retribution or genocide; and these are only the more obvious examples, the tip of the iceberg. If this shocks the reader, which I doubt, then instead of protesting innocence, it might serve humanity far better to individually (the writer included) and collectively acknowledge that this is the truth, or something very close to it and do what we can, however small, to help remedy the situation.

However, all said, there is still much that is good about humanity, much that is lovely, many acts of kindness, many positives to dwell upon.

It has been said that everyone loves a good story, especially one that ends well. Life's story has many mysteries, much intrigue, endless adventure and romance in abundance but love is the glue that makes for a really good ending, or beginning, or an endless series of beginnings and endings. The choice is ours.

A Few References

As a student who has spent many years at different colleges and universities writing essays and theses, I have found that the general reader tends to pay minimal attention to the appended references. The major exceptions are 'examiners' who are mostly very meticulous, some would say pernickety, about the total adequacy, academic merit, format and punctuation involved. Whilst the writer applauds academic rigour, he doesn't subscribe to the idea that a lengthy academic list of references is necessary purely to impress the reader. Given the abundance of online search engines like Google and encyclopaedias like Wikipedia, I believe the general reader is quite capable of researching the authenticity and validity of any particular part of this essay, should they wish to do so.

However, having rather radically applied 'Occam's Razor' to this section, the writer does have a list of suggested books which might be of further interest to the reader.

Three books by Roger Penrose: *The Emperor's New Clothes, Shadows of the Mind and The Road to Reality* and *Cycles of Time.*

One book co-authored by Roger Penrose and Stephen Hawking: *The Nature of Space and Time.*

One book by Stuart R Hameroff: *Towards a Science of Consciousness.*

Several books by Stephen Hawking including: *Black Holes: The Reith Lectures* and *On the Shoulders of Giants.*

One book by Richard Dawkins: *The Selfish Gene.*

Any one of a number of books on Greek Philosophy, Richard E Allen: *Greek Philosophy: Thales to Aristotle.*

Any one of a number of books on the History of Humanity, possibly by David Graeber and David Wengrow: *The Dawn of Everything.*

Last but not least: *The Amplified Bible*, or any other version you may prefer.

What are Stories?

The author's definition of reality is the sum of all the stories told by the lives we lead both individually and collectively, the story of humanity. This definition relegates the material world including our own bodies and the body of all living beings to the stage on which we act out our lives. What then are stories other than forgettable words, some true, some false, words, describing the thoughts, emotions and actions of fleshy creatures given to visions of an eternal life free from sin and death.

Stories are dancing photons which form pictures that tell a story without words. The pictures always stand for the truth; the words we use to describe the pictures are sometimes true sometimes false, often misunderstood.

The duality of light as both photons and waves offer a link to our understanding of reality. Light is a form of energy, and energy and mass are of the same essence. Energy cannot be created or destroyed only changed from one form to another, it is the source of all being and doing and without energy there is nothing.

Photons at rest have zero mass but when they are moving, they are subject to the force of gravity like everything else in the material universe. Black holes swallow spacetime, but black holes still emit energy, dark energy. Soon the dancing photons will appear again in 'different clothes', but will they remember the dance? If photons remember the dance, then stories never die, they are eternal along with all the players in the story.

There is compelling evidence to support the idea that photons do remember, but that is another story.